# INCREDIBLE CHANGES ON EARTH

# LANDFORMS AND HOW THEY ARE MADE

Julie K. Lundgren

## A Crabtree Seedlings Book

CRABTREE
Publishing Company
www.crabtreebooks.com

**Landforms (LAND-formz):** all the natural features of Earth's solid surface

# TABLE OF CONTENTS

# Landforms, Patterns, and Maps

Earth's surface varies. It has natural features of all kinds that make up our planet's **landforms**.

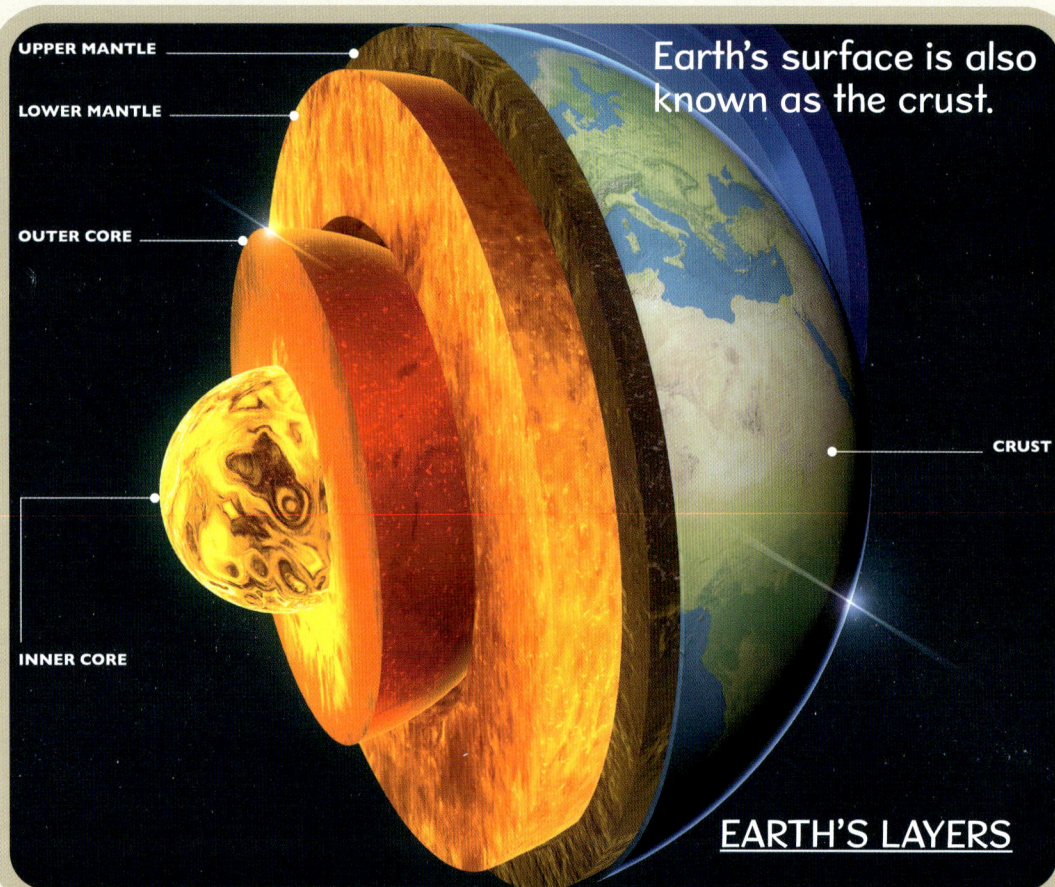

UPPER MANTLE

LOWER MANTLE

OUTER CORE

INNER CORE

Earth's surface is also known as the crust.

CRUST

EARTH'S LAYERS

Some landforms, such as coral reefs, lie underwater.

Great Barrier Reef, Australia

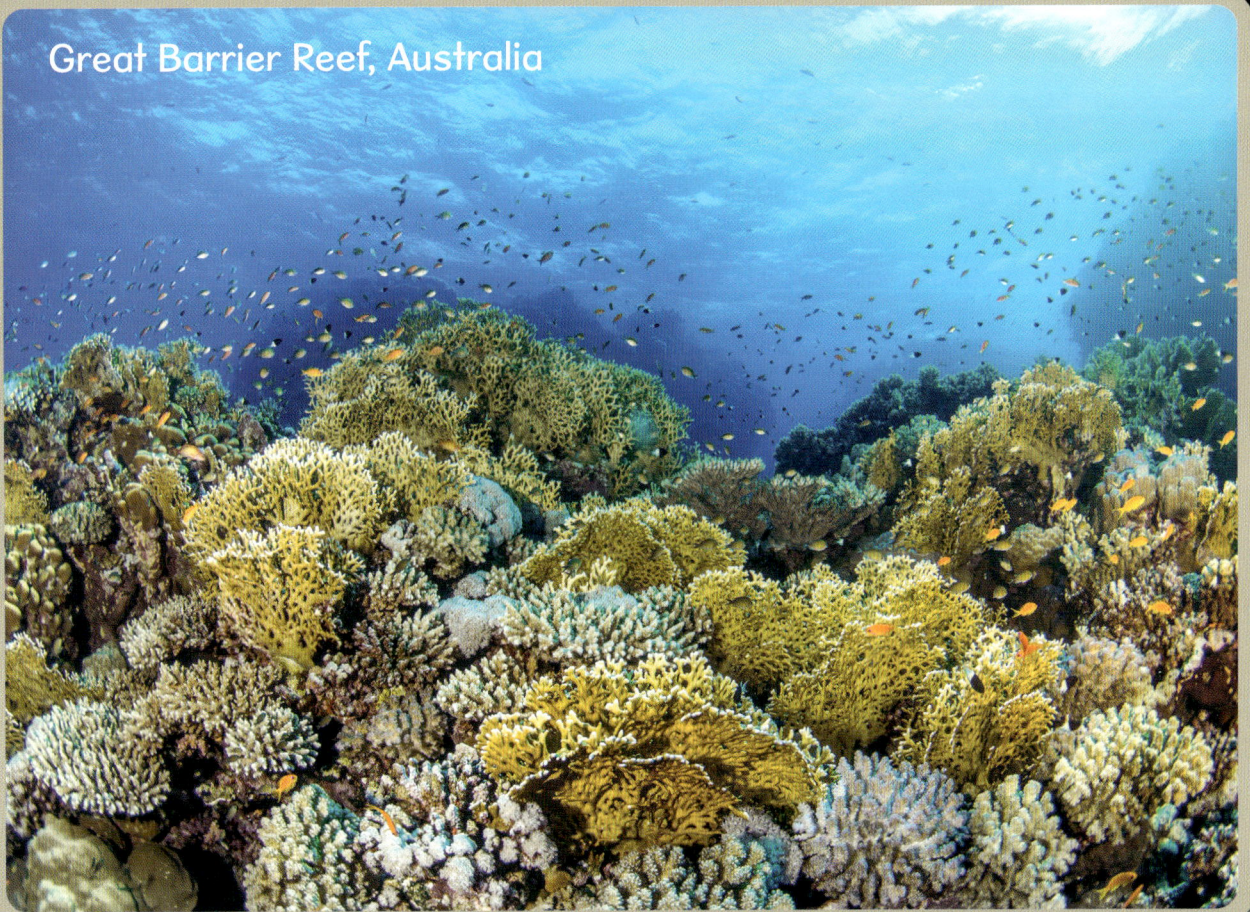

We use maps and **satellite** images to show the shape and location of landforms on the continents and in the oceans.

satellite

Map of the world

Arctic Ocean

North America

Europe

Asia

Pacific Ocean

Pacific Ocean

South America

Africa

Australia

Atlantic Ocean

Indian Ocean

Southern Ocean

Antarctica

The Arctic, Atlantic, Indian, and Pacific Oceans have been named for years. The Southern Ocean, around Antarctica, is our newest named ocean.

Mountains are the tallest landforms.

Landform patterns provide clues about Earth's long history, and the forces at work that make and shape landforms.

Rock layers show Earth's history.

## In The Know!

The **geosphere** is Earth's system having to do with rocks, soils, and minerals.

# Forces at Work: Wind

Wind moves particles of soil, sand, and dust through the air. They brush, hit and scrape surfaces.

Wind carves rock in a desert.

A haboob is a strong wind that creates large dust storms and sandstorms.

## In The Know!

The atmosphere is the system having to do with Earth's air.

Over time, windblown particles wear rocks down so they become smooth. **Deposits** of particles in new places make new landforms.

Wind sweeps sand into dunes.

# Forces at Work: Moving Plates

Earth's crust is made of **tectonic plates**. They fit together like loose, moving puzzle pieces.

Earth's tectonic plates push, pull, and shift.

## Tectonic plate boundaries

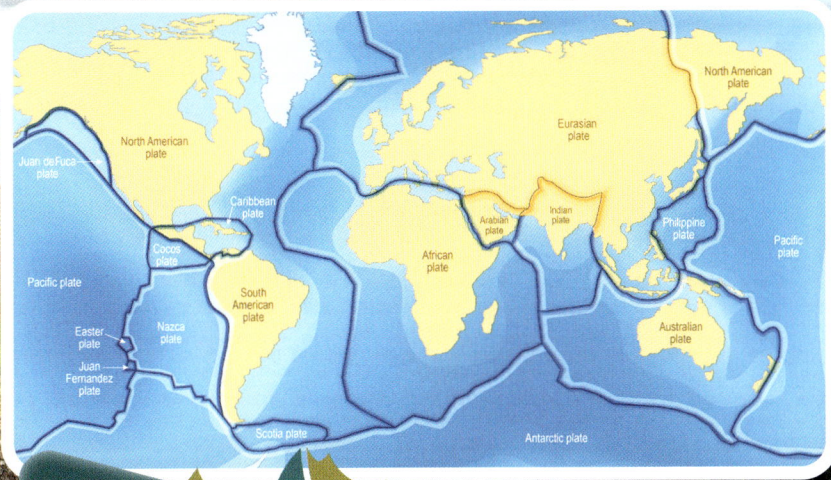

North American plate

Eurasian plate

North American plate

Juan de Fuca plate

Caribbean plate

Arabian plate

Indian plate

Philippine plate

Pacific plate

Cocos plate

African plate

Pacific plate

South American plate

Easter plate

Nazca plate

Australian plate

Juan Fernandez plate

Scotia plate

Antarctic plate

## In The Know!

The **lithosphere** is the system having to do with Earth's outer layer, including Earth's crust.

In Iceland, the North American plate and the Eurasian plate meet.

15

Most earthquakes and volcanoes occur at the edges of tectonic plates. **Magma** rises through volcanoes, making new rock.

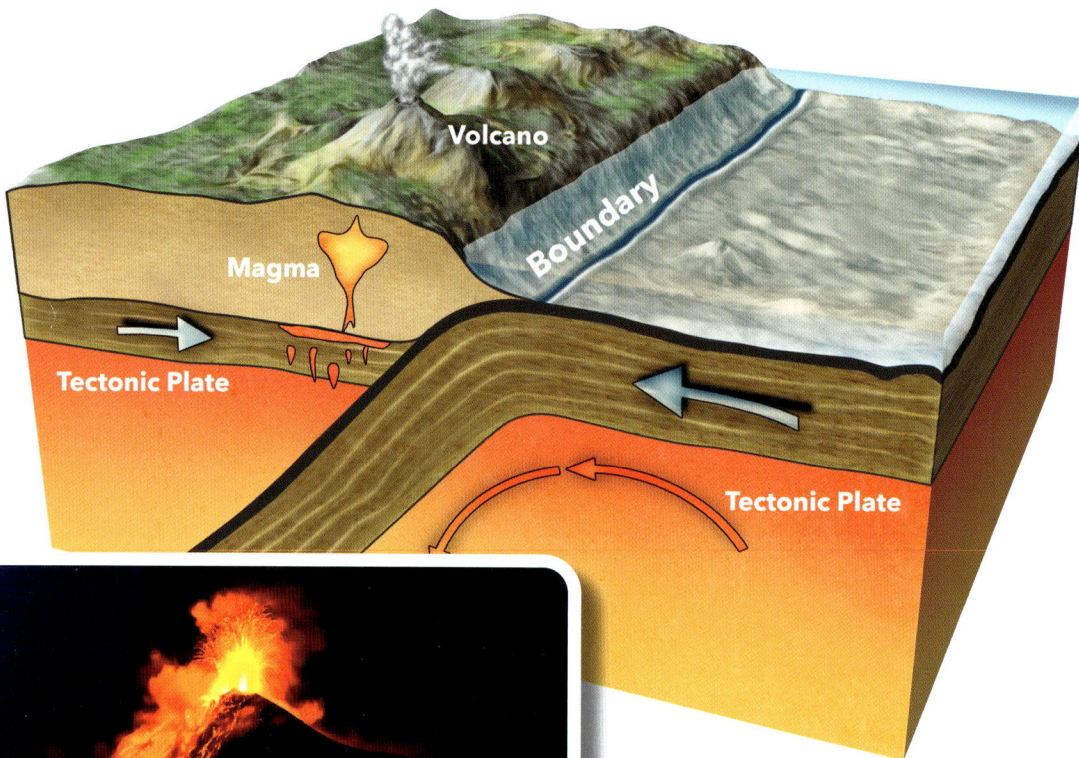

Volcano

Boundary

Magma

Tectonic Plate

Tectonic Plate

Mount Everest—the highest mountain in the world.

Earth's tallest mountains were made when tectonic plates pushed into each other.

# Forces at Work: Water

Water moves soils. Oceans, rivers, lakes, rain, and ice shape landforms through **erosion** and deposits.

## In The Know!

The **hydrosphere** is the system having to do with Earth's water.

Cyprus, an island in the Mediterranean Sea

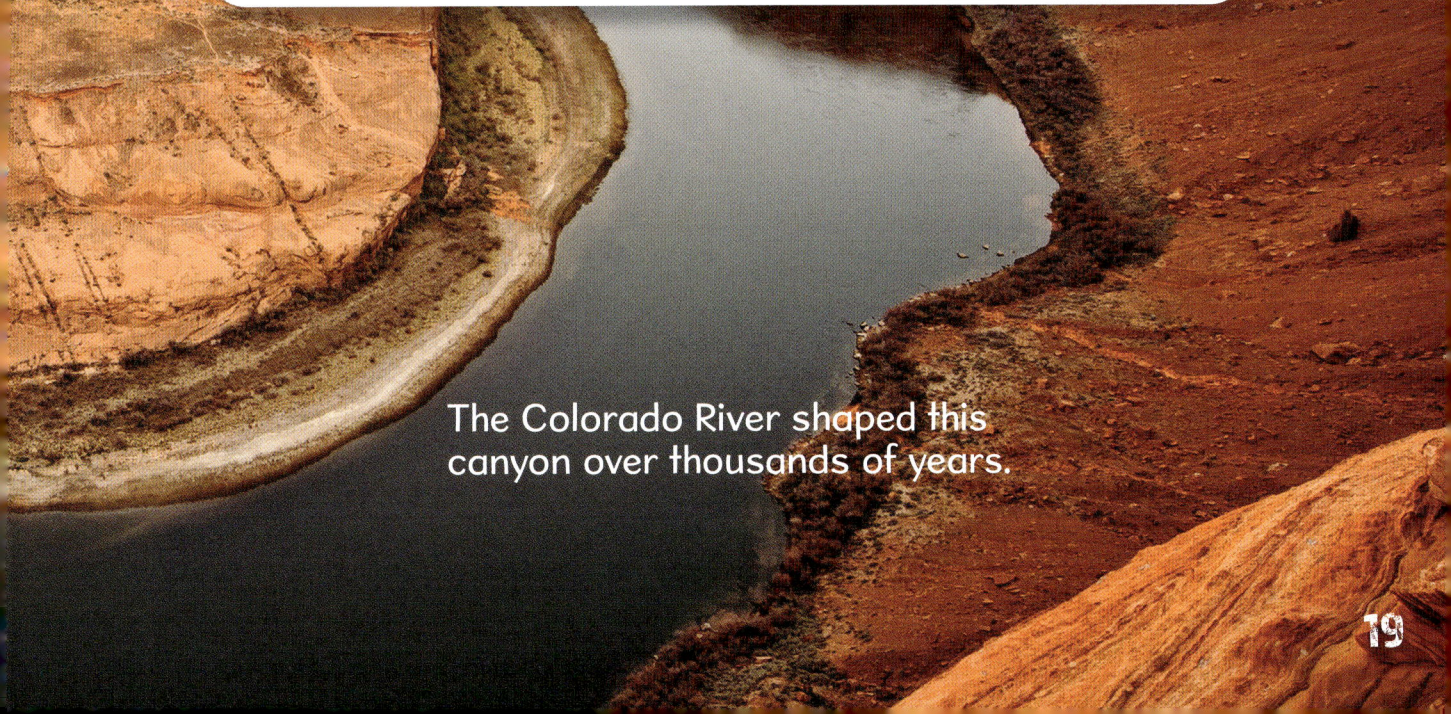

The Colorado River shaped this canyon over thousands of years.

People, plants, and animals interact with landforms and Earth's natural systems.

# Glossary

**deposits** (dee-POZ-its): Layers or piles of sand, soils, or rocks dumped by wind or water

**erosion** (ih-ROH-zhuhn): The wearing away of rock or other surfaces

**geosphere** (JEE-ohs-feer): The solid parts of Earth, including rocks, soils, and minerals

**hydrosphere** (HI-druhs-feer): The system that refers to Earth's water, including rain, oceans, rivers, and lakes

**landforms** (LAND-formz): All the natural features on Earth's solid surface

**lithosphere** (LITH-ohs-feer): The system that includes Earth's crust and its plates

**magma** (MAG-muh): Hot, melted rock below Earth's surface

**satellite** (SAT-uh-lite): Equipment that is sent into space and that moves around Earth, some of which take pictures and gather other information

**tectonic plates** (tek-TON-ik PLAYTS): Vast sections of Earth's crust that change and move

# Index

Climate change, growing cities, and increased mining and industry affect Earth's landforms and systems, and us.

## PROTECT OUR PLANET!

- Reduce, reuse, and recycle.
- Fix leaky faucets.
- Turn off lights when you leave a room.
- Walk, bike, or take the bus.
- Eat foods grown or raised locally.

# School-to-Home Support for Caregivers and Teachers

This book helps children grow by letting them practice reading. Here are a few guiding questions to help the reader build his or her comprehension skills. Possible answers appear here in red.

## Before Reading

- **What do I think this book is about?** I think this book is about rock formations and how they developed over time. I think this book is about physical changes that Earth has gone through.

- **What do I want to learn about this topic?** I want to learn more about the different layers of Earth. I want to learn how water shapes landforms.

## During Reading

- **I wonder why...** I wonder why some mountains are much taller than other mountains. I wonder why and how sand dunes are formed.

- **What have I learned so far?** I have learned that the lithosphere is the system having to do with Earth's crust. I have learned that most earthquakes and volcanoes occur at the edges of tectonic plates.

## After Reading

- **What details did I learn about this topic?** I have learned that Earth's tallest mountains were made when tectonic plates pushed and shifted against each other. I have learned that water moves soils and that rivers shape landforms through erosion and deposits.

- **Read the book again and look for the glossary words.** I see the word *landforms* on page 4, and the word *erosion* on page 18. The other glossary words are found on page 22.

**Library and Archives Canada Cataloguing in Publication**

CIP available at Library and Archives Canada

**Library of Congress Cataloging-in-Publication Data**

CIP available at Library of Congress

## Crabtree Publishing Company
www.crabtreebooks.com        1–800–387–7650

**Written by:** Julie K. Lundgren
**Production coordinator and Prepress technician:** Tammy McGarr
**Print coordinator:** Katherine Berti

Print book version produced jointly with Blue Door Education in 2022

Printed in the U.S.A./CG20210915/012022

PHOTO CREDITS:

www.shutterstock.com, www.istock.com ; Front Cover: shutterstock.com|Dean Fikar. P2-3: istock.com|Nithid. P4-5: Shutterstock.com| Naeblys, Huw Thomas, NASA. P6-7: NASA, shutterstock.com| Rimma Z. P8-9: istock.com|Dmytro Kosmenko, jamesvancouver, shutterstock.com| Jeroen Mikkers. P10-11: istock.com|pawopa3336, shutterstock. com| Caleb Holder, Maxim Petrichuk. P12-13: istock.com|Kim Grosz, Petrichuk. P14-15: shutterstock.com|Mopic, Jose Arcos Aguilar, Designua. P16-17 and back cover: shutterstock.com|Andrea Danti, istock.com|DanielPrudek. P18-19: shutterstock.com| Johnny Adolphson, stock.com|DmitryVPetrenko. P20-21: istock.com|noblige, Sohadiszno, shutterstock.com| Chris Curtis. P22-23: istock.com|chuyu.

**Published in the United States**
**Crabtree Publishing**
347 Fifth Ave.
Suite 1402-145
New York, NY 10016

**Published in Canada**
**Crabtree Publishing**
616 Welland Ave.
St. Catharines, Ontario
L2M 5V6